THE FIELD GUIDE TO GEOLOGY